...CT FILE:
...CHANGED THE WORLD

20 FUN FACTS ABOUT THE LOCOMOTIVE

BY CAITIE MCANENEY

Gareth Stevens PUBLISHING

Please visit our website, www.garethstevens.com. For a free color catalog of all our high-quality books, call toll free 1-800-542-2595 or fax 1-877-542-2596.

Library of Congress Cataloging-in-Publication Data
Names: McAneney, Caitie, author.
Title: 20 fun facts about the locomotive / Caitie McAneney.
Other titles: Twenty fun facts about the locomotive
Description: Buffalo, NY : Gareth Stevens Publishing, [2024] | Series: Fun fact file: inventions that changed the world | Includes index.
Identifiers: LCCN 2022056732 (print) | LCCN 2022056733 (ebook) | ISBN 9781538287682 (library binding) | ISBN 9781538287675 (paperback) | ISBN 9781538287699 (ebook)
Subjects: LCSH: Locomotives–Juvenile literature. | Railroad trains–Juvenile literature. | CYAC: Locomotives. | Railroad trains.
Classification: LCC TJ605.5 .M3993 2024 (print) | LCC TJ605.5 (ebook) | DDC 625.26–dc23/eng/20221230
LC record available at https://lccn.loc.gov/2022056732
LC ebook record available at https://lccn.loc.gov/2022056733

Published in 2024 by
Gareth Stevens Publishing
2544 Clinton Street
Buffalo, NY 14224

Designer: Leslie Taylor
Editor: Therese Shea

Photo credits: Cover, p. 1 (main) Artem Furman/Shutterstock.com; Series Art (file folder) David Smart/Shutterstock.com, (binder clip) luckyraccoon/Shutterstock.com, (wood grain background) ARENA Creative/Shutterstock.com; p. 5 (cargo) Travel mania/Shutterstock.com, (at window) gabczi/Shutterstock.com; p. 6 Hugh Llewelyn/ https://commons.wikimedia.org/wiki/File:Richard_Trevithick_1804_Pen_y_Darren_locomotive_(6635994787).jpg; p. 7 Everett Collection/Alamy.com; p. 8 Everett Collection/Shutterstock.com; p. 9 Jooja/https://commons.wikimedia.org/wiki/File:Steam_locomotive.svg; p. 10 mrubcic/Shutterstock.com; p. 11 E. T. Kenney/https://commons.wikimedia.org/wiki/File:Tom_thumb_peter_coopers_iron_horse_6092027.jpg; p. 12 (last spike) Everett Collection/Shutterstock.com; p. 13 (wagon) ImagePixel/Shutterstock.com, (stamp) MM_photos/Shutterstock.com; p. 14 Public Domain/https://commons.wikimedia.org/wiki/File:US_Civil_War_railway_gun_and_crew.jpg; p. 15 martinova4/Shutterstock.com; p. 16 PTG Dudva/https://commons.wikimedia.org/wiki/File:Number_4468_Mallard_in_York.jpg; p. 17 https://commons.wikimedia.org/wiki/File:Teplovoz_Eel2_(2).jpg; p. 18 The Global Guy/Shutterstock.com; p. 19 cyo bo/Shutterstock.com; p. 20 OlegRi/Shutterstock.com; p. 21 Karasev Viktor/Shutterstock.com; p. 22 Jacob Boomsma/Shutterstock.com; p. 23 (train) Al.geba/Shutterstock.com, (map) Rainer Lesniewski/Shutterstock.com; p. 24 Roelof Nijholt/Shutterstock.com; p. 25 (train) Roberto Sorin/Shutterstock.com; p. 26 cyo bo/Shutterstock.com; p. 27 Michael715/Shutterstock.com; p. 29 andrey_l/Shutterstock.com.

Printed in the United States of America

Some of the images in this book illustrate individuals who are models. The depictions do not imply actual situations or events.

CPSIA compliance information: Batch #CSGS24: For further information contact Gareth Stevens at 1-800-542-2595.

CONTENTS

Words in the glossary appear in **bold** type the first time they are used in the text.

REVOLUTIONARY VEHICLE

Locomotives were some of the most **revolutionary** machines of the 19th and 20th centuries. These **vehicles** were self-propelled. That means they could move using their own power. The first locomotives, made in the early 1800s, were powered by steam.

Locomotives were able to pull train cars full of people as well as **cargo**. Early trains were slow, but they allowed people and cargo to travel at much faster rates than when pulled by horse or other animal. Locomotives shaped history around the world!

train cargo

A train is the name for wheeled cars linked together. A locomotive is the vehicle that pulls it.

THE FIRST STEAM ENGINES

FUN FACT: 1

THE FIRST LOCOMOTIVES TRAVELED ONLY 5 MILES (8 KM) PER HOUR—ABOUT THE SPEED OF AN ELEPHANT WALKING.

British inventor Richard Trevithick built the first steam locomotive that could run on a track in 1804. Pulling five train cars, its weight broke the rails after three trips!

This is a model of one of Richard Trevithick's locomotives.

Some thought early trains could hurt the human brain with their movement and noise.

FUN FACT: 2

PEOPLE WORRIED THAT EARLY LOCOMOTIVES WOULD MAKE THEM ILL.

Steam locomotives were slow. However, people in the 1800s worried about their speed and shaking. They thought passengers, or riders, would be knocked out, unable to breathe, or become ill with "railway madness"!

EARLY TRAIN TECHNOLOGY

FUN FACT: 3

EARLY STEAM LOCOMOTIVES SOMETIMES SET HUGE FIRES ALONG THEIR ROUTES.

Sparks came out of a train's smokestack.

Steam locomotives were powered by burning coal or wood. Grasslands, forests, and buildings along railroads sometimes went up in flames in the 1800s because of stray sparks.

HOW STEAM LOCOMOTIVES WORK

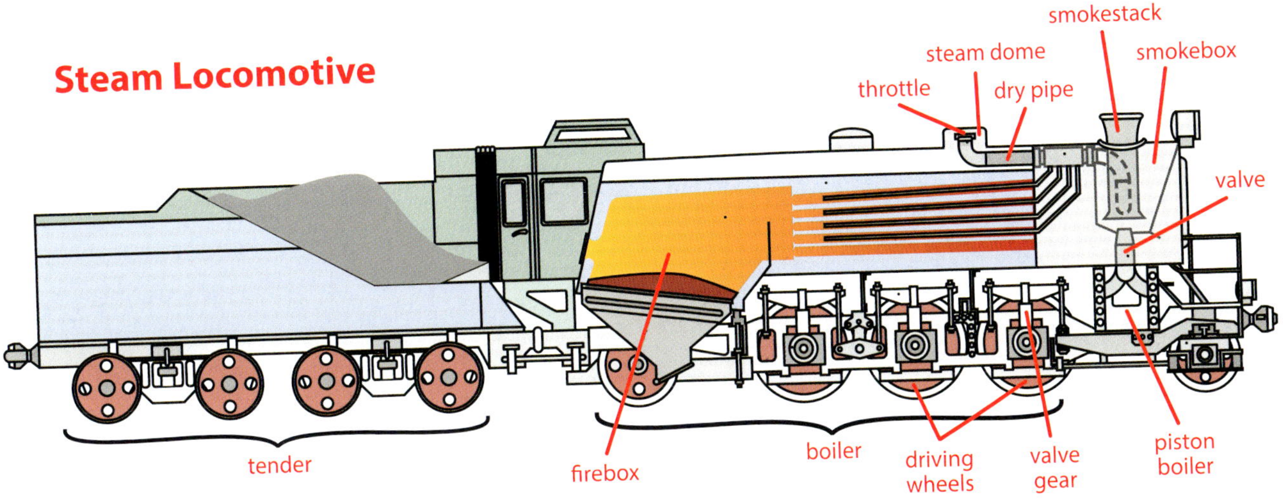

STEP 1:

FIRE IN THE FIREBOX BURNS WOOD OR COALS, LETTING OUT GASES.

STEP 2:

HOT GASES MOVE TO THE BOILER AND HEAT UP WATER, MAKING STEAM.

STEP 3:

STEAM CREATES PRESSURE TO MOVE THE PARTS THAT MOVE THE WHEELS.

This is a basic explanation of the **technology** of a steam locomotive.

FUN FACT: 4

EARLY TRAINS HAD "COWCATCHERS" TO PUSH ANIMALS OFF THE TRACK.

Unlike many British railroads, American tracks didn't have fences. That meant sheep, hogs, cows, and other animals could wander onto the tracks. The cowcatchers helped push the animals out of the way.

cowcatcher

Hitting an animal could cause a train to go off its tracks.

Horse-drawn trains were a standard form of travel until steam locomotives proved successful.

FUN FACT: 5

THE FIRST AMERICAN STEAM LOCOMOTIVE COULDN'T BEAT A HORSE IN A RACE.

In 1830, this first steam locomotive, *Tom Thumb*, was in a testing stage. A horse-drawn train pulled alongside it. The two vehicles had a race. The horse won when a part of the locomotive broke.

THE TRANSCONTINENTAL RAILROAD

FUN FACT: 6

TWO COMPANIES RACED TO BUILD THE UNITED STATES' FIRST TRANSCONTINENTAL RAILROAD.

In the 1860s, the Central Pacific and the Union Pacific railroad companies battled to lay the most tracks. Central Pacific started in the west from California, and Union Pacific started in the east from Iowa.

The companies' railroad tracks met in Promontory Summit, Utah. The new tracks, when connected with tracks in the east, linked the East and West Coasts of the United States.

FUN FACT: 7

LOCOMOTIVES CUT TRANSCONTINENTAL TRAVEL COSTS FROM AROUND $1,000 TO $150.

Many people traveled across the country, some in covered wagons, when workers discovered gold in California in 1848. This was called the California gold rush.

Before the transcontinental railroad, traveling across the United States was costly. It was also dangerous. Some people chose to sail around South America instead. Locomotives made travel safer and more affordable.

MAKING HISTORY

FUN FACT: 8

LOCOMOTIVES HELPED THE UNION WIN THE AMERICAN CIVIL WAR.

The Union had more railroad tracks than the Confederacy. In 1863, about 25,000 Union soldiers were moved south by rail to fight in Chattanooga, Tennessee. The Union won the battle there. This was a **turning point** in the war.

The American Civil War was fought from 1861 to 1865 between the Union (the Northern U.S. states) and the Confederacy (the Southern states). Railroads moved soldiers and supplies.

Making a train **schedule** was nearly impossible before time zones were created.

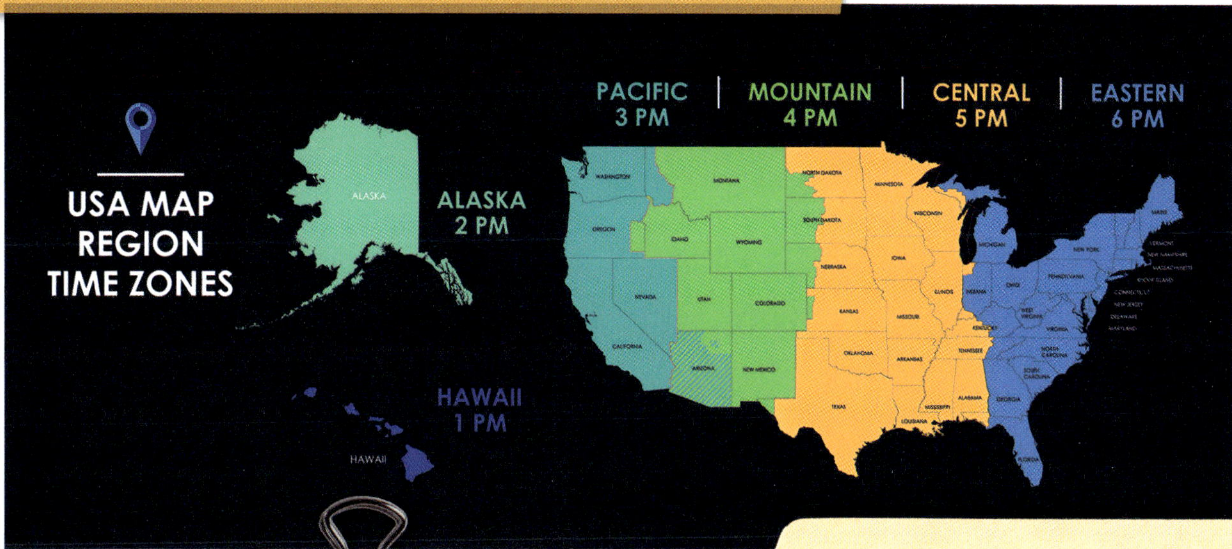

FUN FACT: 9

U.S. TIME ZONES WERE CREATED BECAUSE OF TRAIN TRAVEL.

You can thank locomotives for the different time zones in the United States. Major U.S. railway companies met in 1883 to make time zones across the country. Before that, the time could be different from town to town!

SPEED AND POWER

FUN FACT: 10

THE FASTEST STEAM LOCOMOTIVE TRAVELED ABOUT TWICE THE SPEED LIMIT OF TODAY'S HIGHWAYS!

In 1938, the steam locomotive *Mallard* of the London and North Eastern Railway set a speed record, traveling 126 miles (203 km) per hour. It traveled between London, England, and Edinburgh, Scotland.

The *Mallard* had a sleek and smooth body to help it reach its top speed.

Diesel locomotives came into wide use after World War I (1914–1918). They pollute less and are more powerful than steam locomotives.

FUN FACT: 11

DIESEL LOCOMOTIVE ENGINES CAN MAKE 3,200 HORSEPOWER—ABOUT THE POWER OF 16 CAR ENGINES!

Diesel locomotives use diesel, a kind of thick, oily fuel, to run an engine. The engine powers a **generator** to produce electricity. This electricity runs motors that turn the locomotive's wheels.

FUN FACT: 12

ELECTRICITY POWERS MANY HIGH-SPEED TRAINS AROUND THE WORLD TODAY.

In the late 1800s, electric locomotives were invented. These locomotives use electric motors to make their wheels turn. Some locomotives get their electricity from an electrified track, while others get it through overhead electric cables.

Electric locomotives pollute less than diesel. They're faster too. However, it's costly to build the needed rails or overhead cables.

FASTEST TRAINS IN HISTORY

Fastest Steam Locomotive
- Class A4 4468 *Mallard*, operated by London and North Eastern Railway
- 126 miles (203 km) per hour
- record set in 1938
- United Kingdom

Fastest Diesel Locomotive
- Class 43 power car 43159, operated by Great Western Railway
- 148 miles (238 km) per hour
- record set in 1987
- United Kingdom

Fastest Electric Locomotive
- Class 1216 050, operated by Austrian Federal Railways
- 221.8 miles (357 km) per hour
- record set in 2006
- Germany

Fastest Maglev Train
- Series LO (A07), operated by Central Japan Railway Company
- 374.7 miles (603 km) per hour
- record set in 2015
- Japan

Today's fastest trains are maglev trains, which stands for magnetic **levitation.**

FANTASTIC FREIGHT TRAINS

FUN FACT: 13

AN AUSTRALIAN FREIGHT TRAIN PULLED OVER 682 CARS AT ONCE.

In 2001, the *Mt. Goldsworthy* mining train set the record for heaviest and longest freight train. It weighed about 109,936 tons (99,732 mt) and was more than 4.5 miles (7.2 km) long!

Freight trains carry goods. Some are miles long. The *Mt. Goldsworthy* train carried iron **ore**.

Heavy goods that aren't on a tight deadline are a good fit for freight train travel.

FUN FACT: 14

FREIGHT TRAINS OFTEN NEED MORE THAN ONE LOCOMOTIVE.

Heavy freight cars might carry coal, oil, grains, machinery, or other important cargo. Trains might need locomotives in the front, middle, and rear. The *Mt. Goldsworthy* train used the power of eight locomotives!

CROSSING COUNTRIES

FUN FACT: 15

TODAY, YOU CAN CROSS THE UNITED STATES BY TRAIN IN LESS THAN THREE DAYS.

Since the 1970s, Amtrak has been the main passenger train service in the United States. Passengers can take Amtrak trains from coast to coast. Amtrak runs on over 21,000 miles (33,796 km) of track.

Freight railroad companies own most of the tracks on which Amtrak runs.

The Trans-Siberian Railroad connects key areas of Russia by rail.

FUN FACT: 16

THE TRANS-SIBERIAN RAILROAD COVERS ABOUT ONE-THIRD OF THE WORLD!

The Trans-Siberian Railroad is found in Russia, the world's largest country. The line goes from Moscow in the west to Vladivostok in the east. It takes about seven days to travel its length, about 5,778 miles (9,299 km).

FAMOUS LOCOMOTIVES

FUN FACT: 17

A FAMOUS TRAIN WAS USED IN THE *HARRY POTTER* MOVIES.

Some people ride *The Jacobite* for the view—and some because they're fans of Harry Potter!

Hogwarts Express is a train in the *Harry Potter* series of books and movies. It carries students to the Hogwarts School of Witchcraft and Wizardry. You can ride the train the moviemakers used—*The Jacobite*—through Scotland today.

The *Orient Express* was like a fancy hotel that traveled through Europe and western Asia. Kings, presidents, spies, and wealthy people took this train.

FUN FACT: 18

A FAMOUS MYSTERY STORY IS BASED ON A REAL TRAIN.

In 1883, the *Orient Express* passenger train started running from Paris, France, to Constantinople (today's Istanbul, Turkey). Author Agatha Christie wrote a mystery that takes place on the train called *Murder on the Orient Express*.

TERRIFIC TRAINS TODAY

FUN FACT: 19

MAGLEV TRAINS DON'T TRAVEL *ON* A TRACK. THEY TRAVEL *ABOVE* A TRACK!

Maglev trains can travel nearly 375 miles (603 km) per hour. Magnets push the train upward. Then, magnets push the train forward along a guideway, which is like the train's track.

Maglev trains use powerful **electromagnets**. Maglev trains are used in China, Japan, and South Korea.

Acela trains travel to the cities of Boston, New York, Philadelphia, Baltimore, and Washington, D.C. The newest trains will travel 160 miles (257 km) per hour in certain areas.

FUN FACT: 20

NEW U.S. TRAINS CAN REACH SPEEDS OF UP TO 186 MILES (299 KM) PER HOUR.

As of 2022, the fastest U.S. trains are Acela trains, run by Amtrak. But because of twists and turns along much of their track, they can't travel as fast as they're able to.

THE JOURNEY CONTINUES

Many people think trains are the future of travel. New train technology can be less harmful to nature than cars. Electric or maglev trains could greatly reduce pollution. Trains can move more people at once than buses and cars too.

These new high-tech trains are only possible because of all the different locomotive technology that came before. From the first steam engines to fuel-saving electric trains, locomotives have truly changed the way people travel forever.

Countries such as Japan and China have put a lot of money into making the train technology of the future.

GLOSSARY

cargo: Goods carried by a plane, train, or truck.

electromagnet: A piece of metal that becomes magnetic when an electric current is passed through or near it.

generator: A machine that uses moving parts to produce electrical energy.

horsepower: A unit of power equal in the United States to 746 watts.

levitation: The act of floating in the air.

ore: Matter in the ground from which a valuable metal can be removed.

pressure: A force that pushes on something else.

revolutionary: Causing a great or complete change.

route: A course that people travel.

schedule: A list of times when trains, planes, or buses will leave.

technology: Using science, engineering, and other industries to invent useful tools or to solve problems.

time zone: A geographic area inside which a standard time is used. Earth is divided into 24 time zones.

transcontinental: Extending across a continent.

turning point: A point at which a significant, or large, change occurs.

vehicle: An object that moves people from one place to another.

FOR MORE INFORMATION

BOOKS

Coiley, John. *Train.* New York, NY: DK Publishing, 2022.

Golkar, Golriz. *Eco-trains.* Minneapolis, MN: Bearport Publishing, 2023.

Washburne, Sophie. *Great Train Designs.* New York, NY: Cavendish Square, 2023.

WEBSITES

All About Trains: History and Latest Trends
easyscienceforkids.com/all-about-trains/
Learn more about the history of trains, from the first design to today's technology.

High-Speed Trains
www.dkfindout.com/us/transportation/history-trains/high-speed-trains/
Read a bit about the history of high-speed trains.

Steam Locomotive Facts for Kids
kids.kiddle.co/Steam_locomotive
Explore the history and science behind steam locomotives.

Publisher's note to educators and parents: Our editors have carefully reviewed these websites to ensure that they are suitable for students. Many websites change frequently, however, and we cannot guarantee that a site's future contents will continue to meet our high standards of quality and educational value. Be advised that students should be closely supervised whenever they access the internet.

INDEX